Corn – On and Off the Cob

By Allan Fowler

Consultants
Robert L. Hillerich, Professor Emeritus,
Bowling Green State University, Bowling Green, Ohio;
Consultant, Pinellas County Schools, Florida

Lynne Kepler, Educational Consultant

Fay Robinson, Child Development Specialist

CHILDREN'S PRESS
A Division of Grolier Publishing
Sherman Turnpike
Danbury, Connecticut 06816

Design by Beth Herman Design Associates
Photo Research by Feldman & Associates, Inc.

Library of Congress Cataloging-in-Publication Data

Fowler, Allan.
    Corn — On and Off the Cob / by Allan Fowler.
      p. cm. – (Rookie read-about science)
    ISBN 0-516-06027-9
    1. Corn–Juvenile literature. [1. Corn.]
   I. Title. II. Series.
SB191.MsF657   1994
633.1'5-dc20   94-10471
                                              CIP
                                              AC

What food is grown on more American farms than any other?

Apples? No.
Potatoes? No.
Corn? Yes!

Most of that corn is not
eaten by people.

Farm animals eat a lot
of corn.

The corn that hogs, cattle,
and poultry eat is called
field corn.

The kind that you eat
is sweet corn.

Corn on the cob, fresh
from the farm, is one
way to enjoy it.

But kernels are also taken
off the cob. They are
canned or frozen, so you
can have corn any time
of the year.

One special kind of corn
pops when you heat it.
I'm sure you know what
that kind is called!

There are many other ways to eat corn.

You might eat succotash. That is a mixture of corn and lima beans.

Corn is ground into cornmeal, which is used to make corn bread and corn muffins, tacos and tortillas.

And don't forget cornflakes.

You often eat corn without knowing it. Jam and jelly, candy and soda pop may contain corn syrup or corn sugar.

If these foods make your
fingers sticky, you may
wash your hands with
soap that contains corn oil.

American Indians were the first people to raise corn. They called it maize. We often call it Indian corn.

The Indians taught
the Pilgrims and other
settlers from Europe
how to grow corn.

Corn plants belong to the same family of plants as grass.

Grasses that are used for food — corn and rice, oats and barley, wheat and buckwheat — are known as cereal grains.

A cornstalk usually grows
to be at least 7 feet tall.
Cornstalks in some places
might grow as tall as 20 feet.

On top of each stalk is a tassel. One, or sometimes two, ears of corn grow on each stalk.

The ears are covered by husks. Under the husks are cobs. The cobs are covered with row after row of kernels.

The silky threads on the
ears of corn are called
corn silk.

Most of the corn we eat
has yellow kernels.

But corn can also be
white, red, brown, purple
— or even have different
colors on the same cob.

Farm families used to gather for a party while removing the husks from the ears. They called these parties husking bees.

Today, machines are used to plant, harvest, and husk corn. Machines make it even easier to grow and enjoy corn.

Corn on the cob, popcorn, creamed corn, cornflakes, tortillas, tacos . . .

What is your favorite way to eat corn?

Words You Know

corn

cornstalk

Indian corn (maize)

popcorn

corn on the cob

tassel   corn silk   husk

kernels   tortillas

cornflakes   succotash

Index

About the Author

Allan Fowler is a free-lance writer with a background in advertising. Born in New York, he lives in Chicago now and enjoys traveling.

Photo Credits

North Wind Picture Archives – 17, 26

PhotoEdit – © David Young-Wolff, 8, 11, 15, 24, 30 (bottom right), 31 (center left and bottom right); ©Paul Conklin, 27

Tom Stack & Associates – ©Gerald & Buff Corsi, 23, 31 (top center)

Tony Stone Images – ©Brian Seed, 12, 31 (center right)

SuperStock International, Inc. – Cover, 13, 19, 25, 30 (top right), 31 (bottom left); ©Kro/SuperStock, 7; ©Micheal Rutherford, 9, 30 (bottom left); ©Mick Roessler, 20, 30 (top left); ©Fletcher Collection, 21, 31 (top left), ©Richard Heinzen, 29

Valan – ©Wayne Shiels, 4; ©Kennon Cooke, 14, 22, 31 (top right)

COVER: Corn